This book is dedicated
to all those interested
in understanding the innerworkings
of our plant family.

AF891566

ISBN: 978-0-9981474-3-7

Author:
Lawrence Seals
Illustrator:
Loren Seals

Nitrogen comes to work early
day in and day out.
It is necessary for
BIG TREES
down to little sprouts.

In the PLANT

Nitrogen is needed

to unload all of the cases.

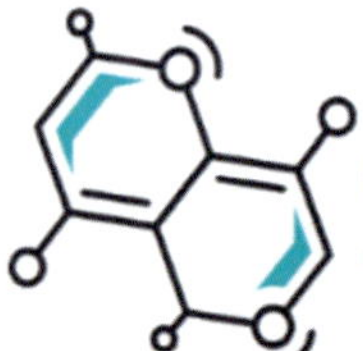

Synthesizing nutrients

and putting things in the right

places.

Before work can get started we must turn on the power. Phosphorus is **PLANT** energy that works for hours and hours.

With Phosphorus in the soil
you just can't miss,
The PLANT is now POWERED
with the flip of a switch!

Now that the PLANT is powered, it's time to get into a groove. Potassium is the supplement that makes everything move.

Moving things around
the PLANT
much like a conveyor.
Potassium can take nutrients

When the cell walls are weak
it can easily show.
Calcium is needed while
the PLANT continues

TO GROW.

A **PLANT** that is in need
can easily break when it bends.
Calcium keeps the stalk strong
to protect from dangerous winds.

As the PLANT gets bigger,
"true" leaves begin to shoot.
The leaves are essential
for turning Sunlight into food.

Magnesium helps leaves
store all the glorious sun rays.
It packs them away
in preparation
for not so sunny days.

If your PLANT is hungry
and in desperate need
of nutrients,
A YELLOWING leaf is a clue
that it most often will present.

Sunlight that was collected
and stored
will certainly be used.
With a little help
Sulfur is able to
distribute the food.

Our PLANT has grown so much
From leaves down to the root!

Healthy, happy, and strong.

It starts to bare fruit.

When all these things are given,
and done just right

The **PLANT** gives you
its *Nutrients*
in every bite.

Made in United States
Orlando, FL
05 November 2026

90224148R00019